SAAB
A SHORT STORY

Brian Williams

For general information on our products and services, please contact us on prodinnova@mail.com

Printed in United States.

ISBN : 9782917260241

10 9 8 7 6 5 4 3 2 1

SAAB
A SHORT STORY

Brian Williams

Contents

Introduction

Forget the "Born From Jets" tag line; it was propellers anyway. Forget the F22 fighter jet used to promote its models, Sweden wasn't involved in its design. Forget reliability issues, we have seen worse in the auto industry. The fact that Saab survived for more than 60 years while building only two platforms: the 92/93/96, and its replacement, the 99 (900) of 1969 (everything since the 900 was begged, borrowed or shared, including with obsolete GM platforms). The fact it never once designed a truly original engine in its entire history, is itself a major feat. This is the more impressive, when you consider the devoted followers and fervent loyal owner-base the company succeeded in creating, and the innovation, the character, the sexy quirkiness it injected into its cars.

It is probably a miracle that such an under-financed company managed to stay afloat until 2012. The fans, who tended to be somewhat offbeat, intellectual individuals (think artists, college professors, industrial designers, architects, engineers, etc.) will certainly miss the ignition switch between the front seats. They will miss the green-glowing instrument cluster. They will miss the low-pressure turbos that give a special feel to Saabs. They will miss the quirky, front-wheel drive which almost nothing else had, extremely safe, very economical, and, last, but hardly least, had design that was truly innovative and different. They will miss the glory days, they will miss the memories of their young time, but their numbers were getting smaller by the day, and the death of the company was inevitable. It was due a long time ago, and it happened in 2012. It has joined a large group of fallen automotive heroes: Panhard, Tatra, Kaiser, TVR, etc. But Saab's case is special. It is a case worth studying. Its' story is worth retelling.

The Beginning

We are at the end of the Second War. In a totally devastated Japan, Toyota was building cars with wooden seats, a single headlight and brakes only on the rear wheels. In relatively tiny Sweden, war wasn't that devastating but Saab was looking for a use for its aviation factories and its engineers. For these companies resourcefulness was essential. They did not have the hundreds of stamping machines that Ford or GM had in the US, they weren't building millions of cars. Saab was not even a car maker. It was not even a civil aircraft maker, although in 1944, seeing the close of war, it began work on non-military aircraft. This is how projects 90 and 91 led to two aircraft: the Saab 90 was a passenger plane seating 25-30. The 91 was a small plane for private use. But project 92 was different. It was dedicated to building a passenger car.

Saab assigned sixteen of its under-employed aviation engineers (of which only two had a driver's license) to this project. Initially, they came up with something a bit more radical and aviation-like, where every opening was covered with a load-bearing hatch. Not practical. So they scoured junkyards, and bought some new cars, including a DKW and an Opel Kadett. The more functional end result was the 92001, or Ursaab. The 92001 prototype was shown to the media on June 10, 1947. It was heavily inspired by the DKW F9, a prototype from the German firm that was first released in 1939 but was never produced in its original form due the War. It also underscored the aviation engineers' relentless pursuit of an even lower drag coefficient. Today, you can still find the 92001 on display at the Saab Car Museum in Trollhättan.

The Ursaab (92001) prototype

The car that ensued, the 92, was Saab's first production car. It was a simple design that grouped all the ideas of chief engineer Gunnar Ljungström and designer Sixten Sason. These included several modern solutions, such as front-wheel drive, transverse

The DKW F9 prototype

engine, torsion bar suspension, rack and pinion steering and a unibody construction with excellent aerodynamics and rigidity. It was Ljungström who opted for rear-hinged doors as he wanted to lessen the risk of damaging the doors whilst driving out of a garage. Saab engineers focused on efficiency and cost as there were rumors that the planned Volvo PV 444 (the 92's home competitor) would be sold at a very low price. It turned out later that it wasn't. When compared to the 92, the PV444 (which was Volvo's all-new car for the post war era), was fundamentally a brick even if it did adopt a bit of hump-backed aero pretense. And built like one too: tough, masculine, conventional in configuration and execution. A solid and reliable burgher. This would be true for all post-war Volvos. There is a saying that perfectly captures their place in the Swedish mindset: Volvo, Villa, Vovve (Volvo, House, Dog). By contrast, Saabs were feminine, creative, intelligent, feline, eccentric, distinctive, and progressive.

***

The unveiling of the Saab 92

The Volvo PV444, not even close to the 92.

Commercial production of the Saab 92 started in December 1949 when a batch of 700 was made. As mentioned in the introduction, none of Saab's engines were original Saab designs. The 92's was borrowed from *DKW*. It was a two-cylinder, water cooled two-stroke, with 764 cc capacity. It was transversely mounted ahead of the front axle and developed 25 horsepower that allowed the car to reach a top speed of 65 miles per hour (105 kph). The transmission had three gears, the first was unsynchronized. The car was already innovative and quirky in many aspects. Never mind the vibrations and the pollution from the two stroke engine which were not unusual at the time. But because of the two stroke design, lubrication was done by the oil that was pre-added to gasoline, meaning the throttle needed to be kept open even when it was not needed. This led to curious situations on slippery roads, where the driver needed to rev the engine and brake at the same time. There were other curiosities as well: Because of their aviation background, the engineers covered the wells of the front wheels (the rear wells were obviously covered) to obtain a body which was aerodynamic as possible, but in snowy Sweden, driving in a straight line, ice accumulated in the wells and rendered the steering inoperative. Often by the time some drivers realized they had no steering, it was too late. But it was not all negative. Saab engineers created a safety cage to protect passengers in case of accidents. The 92 was one of the first cars ever designed for safety. Reinforced pillars remained a feature in all early Saabs.

At the beginning of the 20th century, Ford was painting all model Ts in black, hence the saying: "you can choose any color you like as long as it's black". Ford chose the black color because it dried faster. This allowed him to reduce production cycle times, increase production rates, and reduce cost. Cost was also critical to Saab following the Second World War. 17% of the car was made from imported materials, hence until 1953, you could not get any color for your 92 unless it was green. More precisely, it was a British racing green.

A Saab 92

A Saab 92 with a Saab plane

There weren't any clear explanations for the green color, but according to some sources, Saab had a surplus of green paint from wartime airplane production.

Also, the spare tire on early cars was stored in an externally accessible floor compartment that could be locked from the interior. This compartment was not the trunk. The early 92 had no trunk. There was also an explanation for the absence of the trunk. In order to achieve an aerodynamic profile with a rare Cx (of 0.32) for that time, and to reduce cost and stiffen the rear structure, the top part of the car body, along the rear windshield and rear fenders, were all made in a single piece. This one was made using a gigantic press of more than 90 tonnes and 5 meters height. The press was ordered from a company named *Clearing* and based in Chicago. The giant press was delivered in the summer of 1947 to the port of Göteborg, and then transported on the Göta Älv River all the way to the Trollhättan factory. Its height was so huge that the roof of the body shop had to be raised.

Just two weeks after the release of the 92, Saab entered racing history when Saab's head engineer Rolf Mellde entered the Swedish Rally and came second in his class. Then in 1952, Greta Molander won the "Coupe des Dames" of the Monte Carlo Rally in a 92, tuned to 35 hp (26 kW). It was a delicate foreshadowing of greater things to come.

In 1953, the 92B was introduced with a much larger rear screen and a larger luggage space (with an opening lid this time). It was now available in grey, blue-grey, black and green. In 1954, the Saab 92 power was increased to 28 hp (21 kW) thanks to a new Solex 32BI carburetor and a new ignition coil. In 1955, it acquired an electric fuel pump and square taillights installed in the rear fenders. The colors were grey, maroon and a new color, moss green.

The Amercian Adventure

The 93 was introduced in 1956. It was basically a souped up version of the 92. The engine now had three cylinders and a capacity of 748 cc that allowed it to develop 33 horsepower. But it was still a 2-stroke engine. The front wheels wells were now cut. A freewheel was fitted in order to overcome the problems of oil starvation during engine braking. But the car still had the same three gear gearbox. Then in September, 1957 the 93B was introduced. The original two-piece windshield was replaced with one-piece glass, and two-point seatbelts were introduced as an option. They later became a standard.

A Saab 93

At the same time, Saab had its first foray into genuine sports cars with the Sonett. Developed in a barn by a few enthusiasts, it had a 57.5 hp version of Saab's new three-cylinder two-stroke. Weighing some 1300 lbs, it was a brisk little barchetta good for

100 mph, whiach was nothing to sneeze at in 1956. Racing would have been its purpose in life, had the rules not suddenly changed. Although only a handful were ever built, it was followed by an improved Sonett II and later by a Sonett III.

A first generation Sonett

The 93, which was sold until 1960 before being replaced by the 96, was instrumental to the international expansion of the marque, especially in the US. But Saab's aviation background played a role as well. Indeed, as a successful aircraft manufacturer, Saab already had an international network of parts distributors. In the U.S., Saab's parts-buying agent was Ralph Millet, an ex-pilot, aeronautical engineer and graduate of Massachusetts Institute of Technology. Millet's company, *Independent Aeronautical*, was based in New York and was close to Trollhättan-based Svenska Aeroplan Aktiebolaget (SAAB).

In late 1955, Saab's chairman, Tryggve Holm, went to the US and met with Millet. Although the discussions were dedicated to the sale of aircraft parts, Holm asked Millet, between business discussions, his opinion about importing the new Saab 93 to the US. Millet was pessimistic about the idea and skeptical of American consumers' acceptance of a two-stroke-engine-powered car, as it was necessary to mix oil into the gas tank, like a motorcycle or a lawn mower. At the same time, Millet confessed that, frankly, he knew nothing about the car business.

Two days later, as Millet was driving Holm to the airport, Holm insisted on sending a few cars to be shown at the following major auto show, to see how the public would react. Without delay, five Saabs were shipped, and Millet dutifully booked exhibition space at the 1956 New York Auto Show. Three cars were shown: two Saab 93 models and a Sonett Super Sport.

While one of the two exposed models was a road-ready car, the other was a partially cut-away model revealing the unusual engine, front-wheel-drive and hearty steel construction. The Sonett Super Sport was a limited-production roadster originally intended for competition; only six units were ever built. As an original Saab "concept car," the Sonett Super Sport was a sensation on the auto show circuit.

The public's enthusiasm for the cars helped dissolve Millet's original skepticism. Millet said, "On the first day of the New York Auto Show, I was an expert on spare parts for aircraft. By the final day, I was in the car business with Saab."

He formed a new company, Saab Motors Inc., first as a subsidiary of Independent Aeronautical, and then taken over by the Swedish Saab parent company. Before the end of the year, Millet was president. The new company established its very modest "headquarters" at a small office on West 57th Street, in Manhattan.

The first sizable shipload of 200 93s arrived several months after the New York show, and Millet focused his marketing efforts almost entirely on the Northeast. Fifteen dealerships signed up the first year, and Saab established a warehouse and vehicle preparation facility at the port of Hingham, near Boston.

One of Millet's first promotional activities was to enter three 93s in the Great American Mountain Rally, during Thanksgiving weekend, 1956. With Saab's enthusiastic support, chief engineer Rolf Mellde came from Sweden to drive one of the cars and American rally driver Bob Wehman was recruited to drive another of the entrants.

Fresh snow made the grueling 1,500-mile, three-day winter race even more challenging for the 63 competitors, which included many American brands, as well as Austin-Healey, Renault, Triumph, Volkswagen, MG, Jaguar, Volvo and Mercedes-Benz. After three days of sliding around slick roads with snow up to 16 inches deep, most cars did not finish. Only one American car finished among the top 20 – and to everyone's surprise, first place went to one of the new Saab cars. Wehman piloted a 93 to victory, followed by Mellde in sixth place, while the third Saab finished seventh. Saab took the team award and finished first, third and fourth in its class.

Great publicity accompanied Saab's outstanding performance in the rally, with much credit attributed to Saab's remarkable front-wheel handling, Sweden-bred heater and robust construction. Locally and nationwide, word spread among car enthusiasts about the new import from Sweden. *Road & Track* was impressed enough to note, "The performance has done more to win respect than a million dollars' worth of advertising."

In 1957, the first full year of U.S. sales, 1,410 93s were sold, approximately 14-percent of Trollhättan's total output. By the end of 1959, some 12,000 Saab 93s had been shipped to the U.S., making it Saab's biggest export market.

The two-stroke engine was well suited for winter operation, and owners were extremely satisfied with the fact that it was always willing to start, even on very cold days. Salespeople would promote the fact that there were only seven moving parts to this simple engine: the crankshaft, three pistons and three connecting rods. Besides, the tiny engine fit perfectly in front of the transaxle and its lightness engendered good handling. But original it wasn't; the engine, drive train and the whole layout of the 92 and the 93 were heavily cribbed from DKW, the mother of all *zweitakter*. Also, it was not without flaws. It was smoky, and you couldn't spot an accelerating 93 without a plume of blue smoke behind it. Lubrication problems due to long stretches of consistent-speed highway driving or an incorrect fuel-oil mixture could lead to engine seizure, a catastrophic problem that required the motor to be rebuilt. Rather than ship the broken engines back to Sweden, Millet set up an engine rebuilding workshop at the Connecticut warehouse facility. "We had an assembly line - two or three people - working to rebuild engines," recalled Len Lonnegren, Saab's public relations chief from 1963 until 1989. "Regardless of the problem, it was often best to simply replace the engine - a relatively quick and easy process in an early Saab. We kept many customers quite happy and loyal by doing this without charge, as Ralph Millet had initiated a lifetime engine warranty to boost confidence in the two-stroke engine."

While Saab executives in Sweden were not enthusiastic about this America-only policy, Saab dealers were quick to promote the lifetime warranty, which covered the engine as long as the car belonged to the original owner.

Who were the brave US Saab buyers in the early days? "The customer was generally a detail-oriented, technical person who appreciated the machinery of the car," said Lonnegren. "Many were in an engineering field, or small business owners, or professionals. They were people who read Popular Mechanics. And they were all very enthusiastic."

A survey conducted by Saab in early 1957 revealed that doctors were the largest single group of customers, followed by sales executives and aircraft industry employees. In fourth place was a significantly large group of amateur racing drivers. Another survey, taken almost two years later, classified the largest group as highly educated members of various liberal professions, such as doctors, lawyers, engineers and college professors. Strikingly, how did the same crowd that adopted these smoky cars as a badge of auto-intellectual superiority later become early Prius adopters? Atonement for prior sins?

Saab's international rally heritage was the inspiration for the Granturismo 750, a 93-based special model created primarily for the American market after much persistence from Millet. Introduced at the 1958 New York Auto Show and discontinued in 1960, the GT750 had additional sport-luxury features such as a wood-rim steering wheel, sport seats, driving lights, tachometer and a rally timer, plus twelve more horsepower.

The Grown up Child

The era of the 93 came to an end in 1960, when the new 96 was introduced. It still used the chassis of the 93 (and the 92 before it). This point is important because the production of the 96 ended only in 1980, which means that this chassis (92/93/95/96 family) had been in service for more than 35 years. The 96 itself remained in production for 20 years. Over this period, 730.607 96s were produced.

The 96 can be defined as the ultimate first generation Saab. It started its career with the same three cylinders engine as the 93, but this one was offered with 750 or 841 cc and was good for 38 to 57 horsepower. This was until 1967, when due to emission regulations, a new V4 was introduced. This one had a capacity of 1.5 or 1.7 liters and developed between 55 and 65 horsepower.

Beyond the two-stroke, The 96 offered more trunk space, more glass and a few technological advancements over its predecessors. Swedish rally star Erik Carlsson — aka "Mr. Saab" — got his start in a 93, but he used a 96 to string up a seemingly endless series of wins in the early 1960s. These wins put Saab on the map. The 96 could be defined as a weirder, sportier, more charming Volkswagen Beetle.

Like every Saab before it, the 96 had a column-shift manual with a freewheeling clutch (to address the issue of oil lubrication in a two stroke); when you let off the throttle, the engine was disconnected from the wheels. In 1949, this was a little odd, but in 1980, it was very cool.

A Saab 96

Today it is hard to look to this first generation of cars without being seduced. Old Saabs just have a way of stirring the emotions. You get seduced (forever) by the purity of the line on that svelte fastback, and the expressive headlights. It is clear that the gutsy way with which that bunch of underemployed aeronautical engineers came up with the original 92 was truly inspirational, even if they were heavily inspired by DKW in what was beneath that slippery skin.

***

In the mid sixties, nascent environmental awareness (and the EPA) forced Saab to finally abandon the stinky two-stroke. Finding a suitable four-stroke to fit in the tiny space where the little two-stroke had long resided was no small challenge. But in 1967, the 96 emerged with a four stroke engine. It was the Ford Taunus V4, a 1498 cc engine, originally developed for the 1962 Ford Taunus 12M. Saab's project to source a four-stroke engine was dubbed 'Operation Kajsa'. Four-stroke engines had been tested in Saabs earlier. Between 1962 and 1964 Kjell Knutsson and Ingvar Andersson under Rolf Mellde tested three different engines, Lloyd Arabella 897 cc and 45 hp, a Morris Mini 848 cc, 33 hp engine and a Lancia Appia engine of 1089 cc and 48 hp. However Rolf Mellde's view that Saab needed to switch to a four-stroke engine was stopped higher up by CEO Tryggve Holm. Mellde then went behind the back of Holm and made contact with Marc Wallenberg, son of Marcus Wallenberg, Saab's major stockholder. The coup succeeded and testing could begin. The engines that were considered and tested were the Volvo B18, Ford V4, Triumph 1300, Lancia V4, Opel, Volkswagen and Hillman Imp. The B18 was the most reliable, but the Ford V4 was not far behind and significantly easier to fit into the engine bay of the 96. The testing was done in secrecy. Rolf Mellde took a leave of absence and said he was going to run his father's paint shop. In reality, he went to Desenzano in northern Italy with a 96 V4 prototype for testing. With five months to go before production, only seven people knew about the new engine. To maintain secrecy they rented a house west of Kristinehamn. To keep purchases of V4 specific parts secret they started the company Maskinverktyg AB. The ordinary purchase department at Saab was oblivious to what was going on, something that caused an incident when Rune Ahlberg cancelled the orders for cables for the two-stroke engine and the purchase department called the supplier and sharply told them to keep to their deliveries. In the last week of July 1967, just before the summer holidays, the information about the new engine was given to some more people and they were informed that full scale production would start in four weeks. To keep secrecy, 40 of the ordinary staff were told to report to work to fix a problem with the disc brakes. The secret was kept until a journalist, some days prior to the official introduction, noticed a lorry loaded with 96s that had V4 stickers on the front bumpers.

The Ford V4 was an ultra compact 60 degree engine; it was originally intended for Ford's VW fighter in the late fifties, the aborted Cardinal. The engine was shipped off to Cologne, Germany, where the V4 and its six cylinder offshoot powered millions of Euro-Fords, before finding its way back home into millions of Explorers and Mustangs. Forty years after the first Ford V4 powered Saab coughed to life, Ford's closely related

V6s were still rolling off that same transfer line. Ford certainly amortized this design well.

While the Ford engine was right-sized and right-priced for the 96, it wasn't right-sounding. Whereas a 60 degree V6 has a fairly high degree of harmonic and firing balance, cutting off two cylinders makes a V4 sound and feel exactly like... a V6 with two dead cylinders. A rougher and less even-firing engine would be hard to invent. Ironic too, considering how exceptionally smooth the little two-stroke had been. There was a good reason DKW and Auto-Union used the 3=6 slogan for their 3 cylinder two-strokes: they had the same number of power impulses and the smooth running characteristics of an inline six. The V4 was simply rougher than the two-stroke and nowhere near as goofy. But for the 96, it was just another continuation of its eccentricities: from engine blubbering to engine stuttering.

Nevertheless, the lumpy V4 had a lusty torque curve, good mileage, a clean exhaust, and kept the Saab eccentricity quotient intact. Now I've always thought that the Mazda rotary would have been the perfect successor. It would have fitted like a glove, and a rotary's smooth running and off-throttle popping sounds almost like a two-stroke. It did not happen.

Back to the 96, curiously, despite the introduction of the four stroke engine, the freewheel design was kept until the end of production. In later models, it could be engaged or disengaged by a control in the footwell. Throughout its life span, the Saab 96 and its station wagon sibling, the Saab 95, had the gear lever mounted on the steering column. This became an increasing rarity in the auto industry during the sixties and seventies, but was an appreciated feature among rally drivers who could change gears faster than with a floor-mounted lever. Speaking of the 95, it offered something that the beetle could not offer; It was a seven-seater with a rear-facing folding seat that could be used for children. Packing this level of utility into the 4,300 mm (169.3 in) length was simply a feat that even today cars can't do.

The Sonett re-emerged in 1966, and unlike the original which was a hand-built mid-engine prototype, this time it was a mass produced two-door coupe. It was called the Sonett II and had the US as its intended primary market. Making room for the V4 only challenged its intrinsically compromised lines further.

A Saab 95

It was one of the most eccentric sports cars ever, at least from a mass-producer of automobiles. It was updated as the Sonett III in 1970. The Sonett II proved Saab's engineers, though wacky, were versatile. The V4 was torquey but coarse, and the entire package reeked of kit car. Still, there was no denying the appeal. The attempt to smooth out its bulbous nose on the Sonett III was somewhat successful. It was nimble, had gobs of traction, and essentially ended two inches behind the driver's ass. But with arrivals like the cheaper and infinitely more powerful and handsome Datsun 240Z, the Sonett's few buyers were serious Saabistas, especially since it had only 65 hp, not an eyebrow moving number at that time. Its buyers were certainly not in need of public expressions of their virility.

A Sonett II

A Sonett III

The Age of Maturity

By the mid sixties, Saab was twenty years old, and ready to make its mark in the automotive world. Exit the original platform that started in the forties and that underpinned the 92, 93, and 96. It was time for an ambitious act, and the most defining one; The last truly all-new all-Saab, 99. The 99 arrived in 1967, ready to battle with the likes of the small BMW, Alfa Romeo Giulia, and of course Volvo's also-new 140 Series. It was intended to take the carmaker beyond the market for small cars such as the 96. Despite reflecting a more rectilinear world-view of the times, Sixten Sason's design of the 99 still cut through the air with a very respectable Cd of 0.37. It was roomy, comfortable, handled well, had fine brakes, and offered excellent traction.

A Saab 99

The first prototypes of the 99 were built by cutting a Saab 96 lengthwise and widening it by 7.9 inches (20 centimeters). In order to keep the project secret, the first 99 prototype was badged "daihatsu" as that name could be made up of letters available for other Saab models.

As for the engine, engineering firm Ricardo had assisted Saab in developing its own four stroke engine, but it was going to be too expensive to finalize and put it into production. So Ricardo put Saab in touch with another client, Triumph, which was just about to put its own new SOHC "Slant Four" engine in production. Saab once again did the (seemingly) expedient thing, and had engines shipped from England. It won't come as a surprise to hear that this didn't work out so well. By 1972, Saab started building its own improved version of the engine, now known as the B engine.

Speaking of quirkiness, Saab 99 (and the 900 later) engines were mounted "backwards", with the output and clutch at the front, then feeding power to the transaxle mounted underneath the engine, although with its own oil supply. The 99 was also the first Saab to have its ignition switch between the seats. Unlike other cars, where the steering wheel is locked by the ignition key, the 99 locked the gear stick. The side effect of this was that the driver would always have to park the car with reverse gear activated.

The larger Saab 99 pioneered several Saab world innovations, such as headlight washers/wipers (1970), electrically heated seats (1971), 5-mph self-repairing bumpers (1971) and side-impact door beams (1972). Saab research into active and passive safety systems began with the first Saab prototype, and had intensified ever since. The 99 was the first Saab to be offered in turbocharged form, not to mention one of the first practical uses of such technology in an automobile. It was also the last car designed by Sixten Sason, the Swedish engineer responsible for most of Saab's postwar output and, by the way, the first Hasselblad camera. A 99 was the first turbocharged car to win a WRC event; fittingly, it was also the last Saab the factory took rallying.

In 1974, Saab added a sloping rear hatchback to both its two and four door 99s, creating the combi-coupé in Europe or Wagon Back, in the US. This became a defining characteristic of most Saabs hence, or at least it seemed that way. And it was remarkably roomy back there, thanks to the low floor height, and gave a sporty look to the cars. With a large hatchback door, bumper-height lift-over and fold-down rear seat, Saab's utility set a standard that was followed by other carmakers years later. It was also the closest Saab came in a long time to building an actual wagon at a time when Volvo was churning out brick-shaped wagons by the boatload.

While American cars were losing their mojo during the seventies, Saab's remained very well intact, and continued to grow. The 99 started out reasonably powered by European standards of the time, but that was just a starting point. Increases in displacement, fuel injection, and the sporty EMS model convincingly countered the trend. But the real kicker was the 99 Turbo, which blew a fresh and stiff new breeze upon the automotive landscape. And made indelible impressions on all those who ever got behind its wheel.

At a time when Detroit V8s were making as little as 110 hp, the two-liter turbo four packed all of 145 hp. Sounds ridiculous by today's standards, but in 1978, it was a revelation. Especially when compared to the BMW 320i, which was maxing at 105 hp. It's all relative, and the Saab Turbo helped spark the whole turbo revolution. Soon Dodge Caravans would be proudly sporting turbo badges.

A 99 turbo

***

In the late seventies, safety standards became the focal point of Saab and led to the introduction of 900 in 1979. Because of this, the life span of the Turbo 99 was very short. Very few of these can be found today.

The 900 was loosely based on the 99, which featured a longer sloping hood to comply with the crash test regulations being introduced in the US. Although the 99 could have passed the tests with some modifications, the Swedish automaker felt it was more prudent to bring out a new evolution of the car.

The introduction of the Saab 900 marked the culmination of the most ambitious development ever undertaken by the Swedish automaker. Starting with the tough objective of building a new, larger Saab with high performance and even better road handling, comfort and safety levels than the 99, Saab designers and engineers spent untold man-hours from concept to finished product in developing the new car. Testing of the new model was conducted for several years, in the bitter cold of Scandinavia and the heat of California's Death Valley.

The car was revealed to the press in May 1978, and released for sale the following autumn as a 1979 model year. The 900 re-integrated many of the features of its predecessor including the engine and the styling. It was sold in multiple configurations that used the same chassis. These included the 5 door hatchback, the 2 and 4 door saloons, and the convertible.

The Saab 900

The 900 was 8.4 inches longer than the 99, and boasted a complete new redesign from the front seats forward. The entirely new body was designed by chief designer Björn Envall from the Trollhättan design department. It used a deeply curved front windshield which was far more steep that most cars on the road at the time, calling attention to the marque's aircraft legacy. The hatchback, or "combi coupé" cars were exceptionally spacious. Also underscoring their aircraft lineage, the Saab 900's dashboard was curved to enable easy reach of all controls. Saab engineers placed all controls and gauges in the dashboard according to their frequency of use and/or importance so that the driver needed only to divert his gaze from the road for the shortest possible time and by the smallest angle. This is why the oft-used radio is placed so high in the dashboard. The steering column was telescopically collapsible with deformable steel bellows. In the event of a collision, the steering wheel was drawn away from the driver to minimize the risk of chest and head injuries.

For comfort, a new heating and ventilation system was introduced and, another Saab industry first, a pollen filter to protect the occupants. All models of the Saab 900 were equipped with a low front spoiler made of resilient thermoplastic rubber. The same material was also used as the casing for the famous Saab energy absorbing bumpers. The cellular plastic elements within the bumpers were about 20 percent larger and deeper and have a patented partition designed to raise the bumper's buckling limit.

The 900 was one of the safest cars of its generation

Convertibles, and higher performance models, along with an ever-greater refinement in technology, 16 valve heads, electronic engine controls, and minor body tweaks kept the 900 going all the way to the 26 of March 1993. A remarkable 25 year run for the definitive and longest produced Saab. It became known as the 900 classic or the first generation 900, because, shortly after, Saab introduced a next generation 900 (also known as the 900 II or 900 NG). It was the first Saab to be based on a GM platform (the 1988 Opel Vectra). More about this later.

The 900 cemented Saab's reputation for solid, quick and relatively entertaining cars. It was so likable that it gave the marque what was often claimed to be two decades' worth of brand equity and unstoppable public goodwill (You'll note that we are approaching the 20th anniversary of the end of 900 production). The 900 drove like a small road-going Cessna, or a3-series had BMW made it a front wheel drive. Even base models amble down the road in a distinct, and not entirely unpleasant, trundle. Turbocharged SPG/Aero models went like the devil's polite cousin. And although the 900 wasn't a paragon of reliability, you could change the clutch in less time than it takes to change the water pump, and get less dirty in the process. How's that for logic?

One needs only to watch the movie *Black Cadillac*[1] to understand the cult created around the 900. The movie pitched a 900 Turbo S driven by young college students against a 1957 Cadillac Series 75 Limousine in an epic cat and mouse chase, and at the end the Saab won. The 900 Turbo was also the preferred car of James Bond in most of the John Gardner books, of which *License Renewed* was the first. The 900 was nicknamed the "Silver Beast" and was Bond's personal car. It was equipped by a company called Communication Control Systems, Ltd. (CCS), which did really exist.

***

1 The movie was produced in 2003 and is based on the director's memoirs from the 80s.

Well before the 900's protracted demise, Saab knew it had to be replaced. But the complexities and costs of developing a brand new car were too high. As Fiat was also looking for a partner in order to reduce the costs associated with the design and the introduction of new models, both joined force, which resulted in a project known as Type 4. The explanation was that the platform that would be developed during the project would span four different cars: a Saab, an Alfa Romeo, a Fiat and a Lancia. The cooperation agreement was signed in Early 1979 and work began on the new cars. Alfa Romeo's 164 was styled by Pininfarina while Fiat's Croma, Lancia's Thema and the Saab 9000 were all the work of design studio Giugiaro Italdesign. At first the cooperation went well, but over time differences of opinion between engineers led to friction, and ultimately, very few components were interchangeable between the models. By 1981, the cooperation was simply abandoned. There was a story that says the cooperation stopped when the results of the crash tests were revealed. By Fiat's standards, they were acceptable, by Saab standards, "they were no good at all". The truth is that Saab was focusing on the American market and pre-occupied with the tough American crash tests, while Fiat and Lancia were at the same time withdrawing from the American market, and happy to conform only to the much less stringent European demands. Saab chose to "go it alone" with the Saab 9000. The end product was thus different from its Fiat sisters, despite a certain similarity among the components used. If you take the example of the doors, at first glance, they appeared to be interchangeable between the four cars. However, because Saab fitted heavier side impact protection in their doors, they would not fit in their Fiat sisters.

When the 9000 was first introduced (1984), motoring journals perceived it to be an accomplished executive car with spacious accommodation and admirable build quality. Buyers were sufficiently impressed to tolerate long waiting lists for the new car but this was entirely understandable, as Saab's philosophy is based on iterative refinements of existing designs, rather than copying other makers who see fit to introduce a new model every 4-5 years or so. Indeed, in 1987, one commentator jokingly said that it was very likely that buyers would have to wait patiently until 2001 before Saab would introduce another new design. He wasn't far off the mark with his prediction.

A Saab 9000

A Fiat Croma

One of the curiosities of the 9000 was that it was shorter than the 900, but it offered unequaled internal space. Indeed, the internal space allowed it to satisfy the American EPA criteria for "large cars". The only other imported car that fell into this category was a Rolls Royce. It also offered additional utility with its five-door design (it was designated later as the 9000 CC) which was the only one available at the launch. However, it was a bit challenged in taking on the deeply entrenched and successful mid-sized premium cars like the Mercedes E-Series and BMW 5-Series. Buyers in this class were not so readily moved by the inherent advantages of front wheel drive and a hatchback. A sedan version soon followed, but obviously the front wheel drive was here to stay. The sedan (called the 9000 CS) was launched in 1991 and was produced until 1998.

The "long Run" endurance test on the Saab 9000 at the Alabama International Speedway in Talladega, Alabama, US, in October 1986 played a major role in strengthening the sporty image of the car. Over a period of 20 days, Saab staged an incredible record run with three standard production Saab 9000 Turbo cars. This resulted in 21 new international records and two world records – the foremost of which was a distance of 100 000 km at an average speed of 213.299 km.

The 9000 was the opportunity for Saab to introduce several innovations. In the Saab Direct Ignition system launched in 1988, a separate ignition coil that generated a spark firing 40 000 volts was fitted to each individual spark plug. Three years later, in 1991, Saab was the world's just car manufacturer to introduce a Freon-free air conditioning system.

The usual progression of styling tweaks and performance updates tried to keep the 9000 relevant and attractive. However, the truth was that the 9000 was not a hit, and Saab was in a pickle. The 900 was ageing quickly, and the 9000 was not producing the profits necessary to even contemplate developing a successor cars for either of them. Saab's ambitious push into the premium sector was stalled, and the nose was pointing earthward, precariously so. Time to bail out, or be bailed out. Where were the parachutes?

The Decline

That General Motors would be the one to buy Saab was not a good omen. It was obviously a case of Jaguar envy, after Ford snapped up that equally desperate automaker. Undoubtedly, GM would have preferred to buy something else, something more prestigious, but there were not that many choices. Everyone was getting into the Euro premium car game, and never being one to be left out, GM bit where it could.

GM acquired 50% of money-losing Saab Automobile in March 1990. Thinking didn't appear to be the primary factor; more like fear of getting left behind. That's one of the most powerful decision drivers ever, usually for the worse. And how exactly was GM going to successfully manage another weak brand? At the end of the worst decade of its existence, when its own market share was imploding? In the usual way, by platform sharing.

A 900 NG

Ok, but execution is key, and Saab's (unfortunately named) 900 NG, riding on a 1988 Opel Vectra platform, was quickly seen for what it was: the future of Saab, for better or worse. Saab now had access to capital, technology, and GM's euro engines, but quality and genuine Saab-ness were sorely missing. The NG was introduced in 1994.

Purists would tell you it was not a Saab. GM added some cosmetic changes such as an on-board computer with a display board on the central console (it was called Saab Information Display). Besides including a watch and developing the bad habit of killing pixels, it displayed usual information such as exterior temperature. But more importantly, it also included a new function called "night panel". This function turned off all the instrument cluster lighting except the speedometer (in case of an emergency, related indicators would turn on instantly). According to Saab this was to reduce driver distractions when cruising at night, in addition to playing homage to its aircraft making heritage. Critics wouldn't agree, whilst turning the instrument cluster of a fighter jet off may be useful when one is an American flying in Russian airspace, it was really of little use in a car (Saab was the only one to do it). Critics point out that these are mere gimmicks in the absence of real innovation.

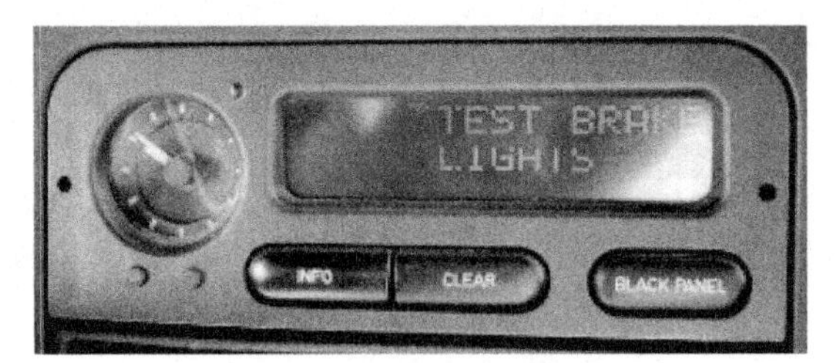

Saab Information Display

After five years of GM's involvement and sanitization, Saab finally showed an operating profit for 1995. It was not to be a regularly recurring feature. Not that it kept GM from buying the rest of the company in 2000; they were too committed by then not to.

Indeed, by then, the replacement of Saab's second best-selling car, the 9000, which was based on another GM platform, the versatile 2900, and which was duly enlarged to accommodate the awkwardly named 9-5, was launched. The 9-5 (pronounced "nine five" rather than "ninety five") was introduced in 1997, but the project (project 640) was started in February 1993 under the direction of Einar Hareide for the overall design and Tony Catignani for the exterior. It was largely inspired by the shape of the Saab 99. For example, the clam-shaped hood covered the front quarter panels and encircled the entire base of the windshield. Additionally the C pillar had the shape of a hockey stick. However, GM increased control and influence over design decisions, suppressing any originality and focusing on cost reduction and component sharing with the Vectra.

The 9-5 family was completed in 1998 with a station wagon. It was also the period when modern diesels (that use high-pressure common rail injection) were all the rage in Europe (in some countries such as France and Austria diesels represented more than 60% of sales). Saab, like Honda and Nissan, were caught without any diesel offering. But GM had Saab covered. And an Isuzu V6 developing 176 horse power was used in the rescue. The engine was rough and far behind the best. It was a first sign of things to come. There were other signs too. Usually, automakers redesign their models entirely every five years or so. Not the 9-5. Introduced in June 1997, it was first face-lifted in 2001 and then a second time in 2005. One of the most controversial changes that were made at the time was the chrome circles around the headlights. These were the idea of Anthony Lo, Director of the Advanced Design Center of GM Europe. These gave the impression that GM couldn't care less about Saab; that designers were running out of options. Yet, some loyal customers bought it. Actually, they were so fanatic that they bought whatever Saab (actually GM) threw at them, and I have seen many of these overly redesigned Saabs on the road.

An overly re-designed 9-5

At the same time, a new 9-3 was introduced in 2002. It was again based on the Vectra platform, which was called the global GM Epsilon platform, and has been lengthened to accommodate four new cousins, the Chevrolet Malibu, the Pontiac G6, and the Saturn Aura. The 9-3 remained in production until 2012.

The 9-3, just like 9-5, had its virtues and vices, lovers and haters, but what can't be argued is that both failed to save the brand. But this was hardly surprising. Merging upscale brands with mainstream ones and having them share platforms never worked. History is littered with such examples. Daimler failed in its merger with Chrysler and left it clinically dead (the 300C and class E shared the same platform), it failed also with Mitsubishi. BMW also failed with Rover and offered it for a symbolic pound to a bunch of British investors. Even VW is failing with its Japanese partner Suzuki. Some say that the Germans, victims of their superiority syndrome, fail to cooperate with others, but who said that Detroit could have done any better.

The final 9-3

Still the truth has to be said; GM with its bean-counters mismanaged Saab. The new-generation Saab lineup of 900 and 9-5 (also, sadly, based on an old Opel) didn't make the cut from the beginning. A more active corporate custodian would have noticed this and taken swift action. GM, however, apparently felt itself to be in the position of a new boyfriend demanding to be serviced in an identical fashion to the old. The 99/900 had lasted twenty-four years and sold well from start to end, therefore the new-gen cars would also have an extended model run regardless of the consequences. The 900 was face-lifted into the 9-3 and rotted in the dealerships for a decade before being replaced by another Opel-platform mediocrity. Just for the sake of perspective, it should be noted that the 900/9-3 was sold against not one, not two, but three generations of Lexus ES, any and all of which were more reliable, comfortable, and practical than the ageing Swede. Even staid old Mercedes-Benz managed to field two new C-Class models during the 900/9-3's extended run.

***

Among the many sayings attributed to Abraham Lincoln is this one: "If I should call a pig's tail a leg, how many legs would it have? Only four, because my calling the tail a leg would not make it so." Although in truth Lincoln merely appropriated this bit of folk wisdom for his own speechifying purposes, it's still relevant when considering some of Saab's models from the new century. They were called Saabs, but they weren't Saabs. Indeed, with a drought of new models, and Saab's two models getting old (in 2004, the 9-5 was already 7 years old), and with losses pilling up, GM resorted to an old trick in the industry called Badge Engineering. GM calls it brand extension. Badge engineering has the benefits of saving resources and reducing costs and could deliver an uptick in volumes in the short term but it inevitably dilutes the brand and kills it in the long term. The first to undergo the badge change was the Subaru Impreza which became the Saab 9-2X. The X was the for the Subaru signature Systematical All Wheel Drive. What it was doing under the chassis of a Saab remains a puzzle, and so was the boxer engine. Not to mention that it was assembled in Japan. Connoisseurs called it the Saabaru. GM did not bother to sell it in Europe fearing it would alienate loyal Saab customers.

That was just the warm-up act. The headliner was the 9-7X built from the Chevrolet TrailBlazer. A Saab born from truck frames and V8s. Probably the best SUV of its kind GM ever built; it was dubbed the TrollBlazer by the critics. GM engineers rushed the vehicle into production so fast that they did not get time to move the ignition switch to the area between the seats. Poor Saab, now undergoing a sex change operation in its old age. What next? GM thought that its upscale Cadillac brand may provide better synergies with Saab. In order to make a better use of the available capacity in Trollhättan, they quickly redid the metal sheet of the 9-3 with sharp angles, et voilà, the Cadillac BLS was born. It was a flop. It was so un-Cadillac that GM didn't have the courage to sell it in the US. But GM did not give up just yet. The Cadillac SRX was transformed into the 9-4X. Built on GM's "Theta Premium", assembled in Mexico, and sold in America with minimal success, there was nothing Swedish or Saabish in it.

The Saab 9-2X

The 9-7X

The Cadillac BLS

The 9-4X

The final car to be introduced was the 2010 9-5. The Opel Vectra platform was still there, as was the green instrument cluster.

The ignition was still between the seats. Actually it is a "start engine" button by now. As usual, the engines were outsourced. But the car was a flop. Some critics argue that the biggest mistake ever made by the company was the introduction of new 9-5 when the 9-3 was ageing. As a two model company, you could survive if one of them was selling well, but you cannot survive when both are slow selling. Alfa Romeo faced the same situation in the mid-nineties, when it was developing the 156 and the larger 166. Initially the 166 was planned to be launched first, but because of financial difficulties, Alfa stopped work altogether on the 166, developed the 156, introduced it successfully and then moved to the development of the 166. This simple decision, inoffensive as it seemed, was a key to the survival of the brand.

The final Saab car (the 2010 9-5)

The End

By the middle of last decade, GM financial situation had deteriorated so much and it became clear that it could not maintain all the brands and participations it had around the world. It sold its Subaru stake to Toyota in 2005. Then in 2006, it sold its stake in Japanese carmaker Isuzu to a Japanese investor. Shortly after, the worst recession since the great recession of the thirties hit the world economy. People stopped buying cars altogether. By 2009, GM filed for bankruptcy protection, and could have disappeared altogether if it had not been bailed out by the US government. In the same year, it killed its Saturn and Pontiac brands. It was clear that GM would not be able to maintain Saab, a loss-making business that lacked the economies of scale and had become irrelevant, for too much longer.

It is interesting to note that Saab fans, with virtually no exception, will tell you that GM "ruined" Saab, but the truth is that Saab would have gone in the eighties had GM not rescued it. But history cannot be rewritten, and Saab is itself a detail in industrial history anyway. They would have been much happier to see Saab go to its inevitable grave twenty years ago, without the GM years and the recent histrionics. Death is never a pretty thing, car companies included. We might have spent the past twenty years arm-chairing endless "what -ifs" and "could-have" scenarios. But it is hard to imagine anyone coming up with a more bizarre outcome.

But as it is always the case, the reality is far more complex than the seemingly bad decisions of a parent company. It requires a thorough analysis. First, as a company, Saab had a product-development timetable that might best be described as "leisurely". This did not matter in the early days, when the best carmaker in the world only replaced their mainline sedans every nine or ten years. But with the emergence of the Japanese in the early 80s and their meticulous, 5-years or so, product cycles things totally changed. Then, the nineties saw every Japanese carmaker introducing a luxury brand. There were Lexus (Toyota), Infiniti (Nissan), and Acura (Honda). Heck, even Mazda was considering one (the Amati). Suddenly the Europeans, which were selling their cars at outrageous prices because of their scarcity, weren't the only kids on the block. The best example of this game changer was the arrival of the second-generation Lexus ES. Based on the 1992 Toyota Camry, arguably the best family sedan in history, the ES300 was flawlessly assembled, impressively equipped, priced in an absolutely predatory fashion, and backed by a monstrous armada of pretentious yet effective marketing aimed directly at the heart of America's nouveau riche. Forget the wooden-seated Toyotas from the end of the War. The Japanese had obviously done their homework. The tweed-jacket crowd didn't cotton to the snub-nosed Lexus immediately — darling, it looks cheap and common— but as tales of the super-Toyota's relentless reliability circulated through the dusty, crowded Saab service-department waiting rooms, surely more than one assistant dean seriously considered the idea of switching loyalties. Most importantly, the 1992 ES was modern, based as it was on a new-for-1992 car. The 1992 Saab 900 was based on the 1968 Saab 99, and it didn't take too perceptive an eye to see it. Of course, by then, Saab had already fallen into the orbit of General Motors, and GM had new products coming. Sort of. The 1993 Saab 900 was based on a 1988 Opel, said Opel being not a very good car. In Sweden, where nobody expected Saabs to be world-beating luxury superstars, it wasn't such a big deal. In America, the press and the public measured it against the competition, ranging from the aforementioned ES300 to the spectacular new E36 BMW, and found it to be well below par.

The Saab story includes airplanes, rally drivers, turbochargers, iconoclastic personalities, and more than half a century of fabulous designs. But the truth is that Saab has been a fake for nearly twenty years, selling second-rate GM cars on dimly remembered glories. Meanwhile, Japanese and German luxury brands have been continually building the cars their customers want, always fresh, nearly always reliable, always sold and serviced with a smile. Saab's better future was perpetually around the corner; meanwhile, the next Lexus or Acura was completed on time and plopped, Harvest-Gold-colored, on a calmly rotating showroom turntable. Ask any Saab enthusiast about the brand and they will tell you about the 900 Turbo S, but ask a Lexus car owner about his car and he will tell you he likes it. Whose is real, and what is fake?

***

By January 2010, GM finally managed to find a buyer for Saab in the form of Spyker, a Dutch manufacturer of supercars with little experience making mass produced cars. By then, the press was already running articles about the demise of Saab. Potential buyers started worrying about service and maintenance while cash-strapped Saab could not advertise the opposite. Saab was already in a death spiral. It had sold 102,915 passenger vehicles in 2007, but only 32,048 cars globally in all of 2010.

In December 2011, Victor Muller, Saab's chairman and CEO addressed the 3400 employees at Saab's Trollhättan plant; they unexpectedly clapped and cheered. Muller, who felt he had failed them, was overcome by emotion. But GM blocked a rescue plan in which Chinese manufacturer Zhejiang Youngman Lotus Automobile would become Saab's largest investor. GM didn't just balk. It flatly refused to approve the vital license transfers—citing China's dismal record of protecting patents and copyrights—which could have put GM in the position of competing with its own intellectual property. Without GM's approval, the licensing agreements for the Saab 9-3, 9-5, and 9-4X were expected to be terminated. And without those cars, as Muller points out, any new investor is "buying an empty factory". He described GM's abandonment as "basically the last nail in the coffin of this beautiful company," before adding "We did the best we could. Now, the only thing that people will remember is that we failed."

Still, Saab's former owners and unrepentant fans of the old cars will always dream of a Saab comeback. That some Swedish folk could create a world-class product, à la Steve Jobs, and humiliate the Japanese juggernauts on the open road. That a stunning new car that had the spirit of that old 99 Turbo and brought the old virtues to a generation not even alive when the only two turbo cars on the market were the Saab and the Porsche 930. But these will remain just dreams. The reality is that the Japanese, and soon the Koreans, have taken over the automobile world, and they are here to stay. Saab is just another casualty in this evolution. Saab is, inexorably and completely, dead.

What a future retro 92 might have looked like

References:

(1) Wired.com, Börn from Jets: The Five Most Awesome Saabs Ever, December 2011.

(2) Saab: The Eulogy, Paul Niedermeyer, ttac.com, December 2011.

(3) My Saab Story, Stein Leikanger, ttac.com, February 2009.

(4) Avoidable Contact: Lexus killed Saab, but GM let Saab die, Jack Baruth, ttac.com, February 2012.

(5) Obituary: Saab Automobile, 1947-2011, David Gluckman, caranddriver.com, December 2011.

(6) Saab is Actually Dead This Time, To Be Sold Off in Pieces, Justin Berkowitz, caranddriver.com, December 2011.

(7) Saab Agonistes: John Phillips on the Death of the Swedish Brand, John Phillips, caranddriver.com, February 2012.

(8) GM Pulls Plug on 9-4X Crossover Production as Saab Moves Toward Chinese Ownership, Justin Berkowitz, November 2011.

(9) Wikipedia Saab 99 page, http//en.wikipedia.org/wiki/Saab_99

(10) Wikipedia Saab 92 page, http//en.wikipedia.org/wiki/Saab_92

(11) Wikipedia Saab 93 page, http//en.wikipedia.org/wiki/Saab_93

(12) Wikipedia ursaab webpage: http://en.wikipedia.org/wiki/Ursaab

(13) Photos from Aftonbladet newspaper website, at http://www.aftonbladet.se/bil/article6917441.ab

ISBN : 9782917260241

Printed in Great Britain
by Amazon